The INSIDE GUIDE

SPACE SCIENCE

The Sun

By Judy Thorpe

Cavendish Square

Published in 2024 by Cavendish Square Publishing, LLC
2544 Clinton Street Buffalo, NY 14224

First Edition

Website: cavendishsq.com

Cataloging-in-Publication Data

Names: Thorpe, Judy.
Title: The sun / Judy Thorpe.
Description: Buffalo, New York : Cavendish Square Publishing, 2024. | Series: The inside guide: space science | Includes glossary and index.
Identifiers: ISBN 9781502670243 (pbk.) | ISBN 9781502670250 (library bound) | ISBN 9781502670267 (ebook)
Subjects: LCSH: Sun–Juvenile literature.
Classification: LCC QB521.5 T56 2024 | DDC 523.7–dc23

Editor: Jennifer Lombardo
Copyeditor: Jill Keppeler
Designer: Deanna Lepovich

The photographs in this book are used by permission and through the courtesy of: Cover Vadim Sadovski/Shutterstock.com; p. 4 Kakoli Dey/Shutterstock.com; p. 6 Nor Gal/Shutterstock.com; p. 7 VICTOR TORRES/Shutterstock.com; p. 8 udaix/Shutterstock.com; p. 9 IgorZh/Shutterstock.com; p. 10 Traveller Martin/Shutterstock.com; p. 12 NASA/SDO; p. 13 magnetix/Shutterstock.com; p. 14 BlueRingMedia/Shutterstock.com; p. 15 Marjan Apostolovic/Shutterstock.com; p. 16 Nazarii_Neshcherenskyi/Shutterstock.com; p. 18 electra/Shutterstock.com; p. 19 VectorMine/Shutterstock.com; p. 21 (main) Andrew Mayovskyy/Shutterstock.com; p. 21 (inset) dreibirnen/Shutterstock.com; p. 22 GizemG/Shutterstock.com; p. 24 MichaelTaylor/Shutterstock.com; p. 25 Macrovector/Shutterstock.com; p. 27 NASA images/Shutterstock.com; p. 28 (top) Johan Swanepoel/Shutterstock.com; p. 28 (bottom) joshimerbin/Shutterstock.com; p. 29 (top) Ground Picture/Shutterstock.com; p. 29 (bottom) Elena11/Shutterstock.com.

CPSIA compliance information: Batch #CWCSQ24: For further information contact Cavendish Square Publishing LLC at 1-877-980-4450.

Printed in the United States of America

The wheels on this ancient Indian temple stand for the wheels on the chariot of Surya, the Hindu sun god. The sun often played a big part in ancient belief systems.

CONTENTS

HUMANS AND THE SUN

Chapter One

Humans have studied the star we call the sun since their earliest days. Many ancient **cultures** had myths, or stories, about it. Some believed the sun was pulled across the sky by a chariot, a ship, or animals such as fish or snakes. Many thought they needed to do some kind of magic to make the sun rise in the morning. In some places, people believed the sun or a god that controlled the sun wanted them to kill an animal or a person to make it rise.

Making Guesses

In the past, before **technology** such as telescopes was invented, people had to guess at why things in nature were the way they were. Sometimes they were right, but often they were wrong. For example, the ancient Greeks believed the sun had healing powers. In one way, they were right! The sun gives off a kind of light called ultraviolet (UV) light that can kill some kinds of germs. However, UV light can't heal all sicknesses—

> **Fast Fact**
>
> In early cultures, people who figured out the movements of the sun and moon were often considered magical. Today, we know this can be done with math and science.

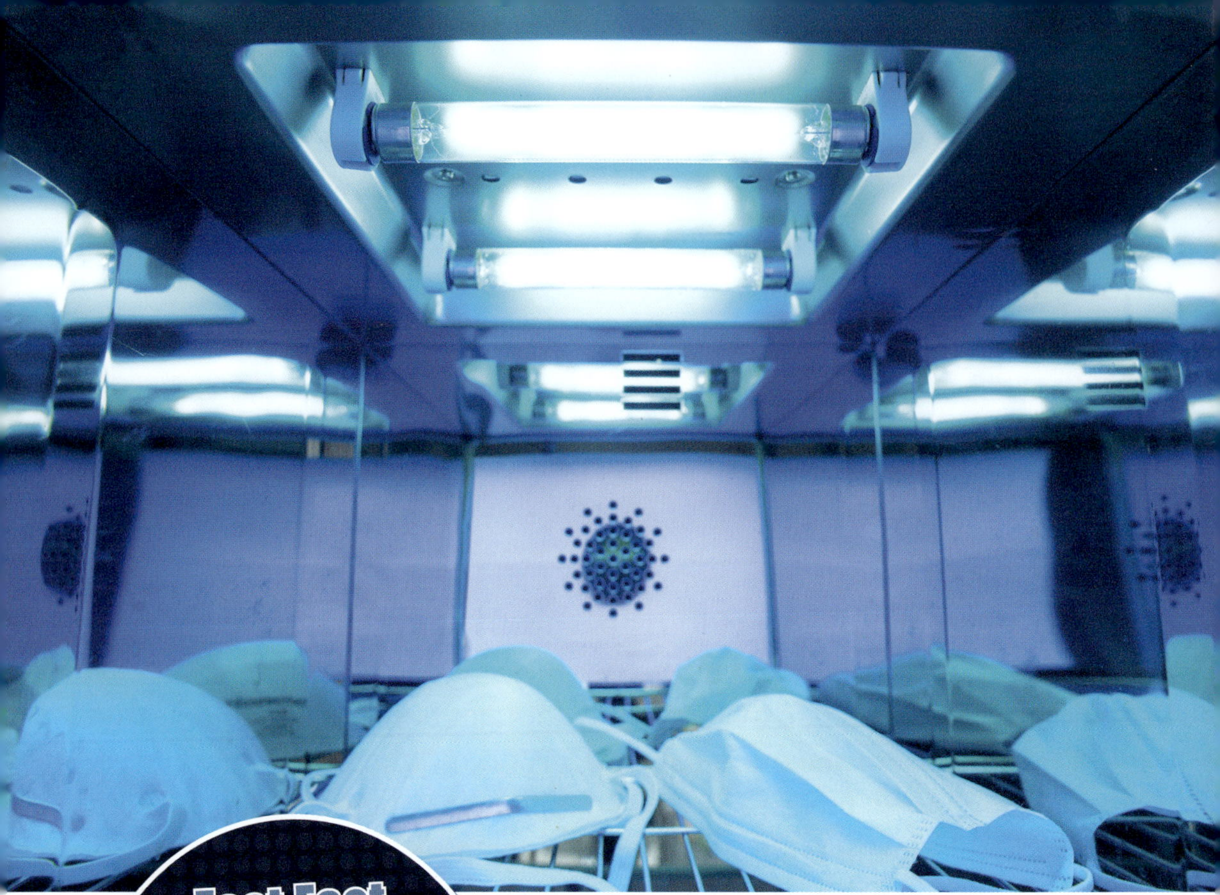

Hospitals use UV light to kill the germs on some items that need to be reused.

Fast Fact

A Greek man named Anaxagoras suggested that the sun was a star in 450 BCE. This wasn't proven until the mid-1800s.

and it also causes some, such as skin **cancer**.

The more technology humans invent, the more we learn. The telescope helped solve some mysteries, but it also created more questions. Early **astronomers** could see dark spots on the sun that couldn't be seen without a telescope. Some believed these sunspots were proof that there were volcanoes on the sun. Others thought they were shadows cast by the planets as they orbited, or circled, the sun. Over time, new tools showed scientists that sunspots are cooler areas where there is a lot of magnetic activity.

Measuring Time

Early cultures quickly figured out that the sun moves in **predictable** ways. They knew that it would shine in certain spots at certain times of the year. They knew that every few years, there would be an eclipse.

Fast Fact

In 1532, a Polish scientist named Nicolaus Copernicus became the first person to suggest that Earth orbits the sun instead of the other way around.

The ancient Maya studied the sun and built this pyramid in what is now Chichén Itzá, Mexico. Every year around March 21 and September 21, a shadow that looks like a snake creeping down the side can be seen.

SEASONS, SOLSTICES, AND EQUINOXES

Planets and stars rotate, or spin, on an imaginary line called their axis. Earth is tilted on its axis, which is why we have changing seasons. When the North Pole is tilted toward the sun, it's summer in the Northern Hemisphere, or half of the planet, and winter in the Southern Hemisphere. When it's tilted away from the sun, the opposite happens.

Solstices and equinoxes mark the beginning of each season. The winter solstice marks each hemisphere's least amount of sunlight in a day, and the summer solstice marks its greatest amount. The equinoxes happen in spring and fall, and they're the two days of the year when day and night are almost the same length.

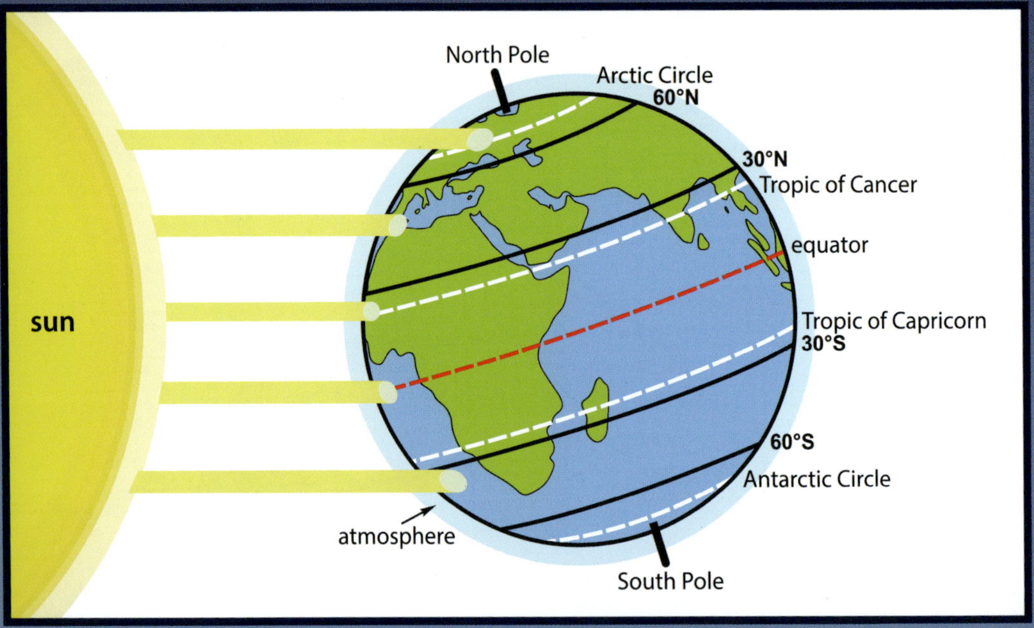

This picture shows how the sun's rays hit Earth at different angles.

People of many ancient cultures built things to help them study the movements of the sun. For example, the city of Machu Picchu was built hundreds of years ago in what is now Peru. It is in the Andes Mountains because it was built to help people study astronomy.

Understanding and predicting the movements of the sun helped ancient cultures make decisions such as when to plant and harvest crops. Many also believed that the sun had a lot to do with luck, so people used it to decide things such as when to travel, when to get married, and when to make business deals. For example, eclipses were believed to be bad luck, so many people avoided traveling or doing anything that was likely to be unsafe until the eclipse had passed.

A solar eclipse happens when the moon passes between Earth and the sun.

We can only see other stars at night because during the day, the sun outshines all of them.

Chapter Two

OUR NEAREST STAR

Like all stars, the sun is made mostly of hydrogen and helium. These are gases that create fuel for the star. They burn in the core, or center, of the star. This makes the star give off light and heat. Without the light and heat from our sun, there would be no life on Earth.

Big and Bright

When we see the sun in the sky, it looks very small. However, this is only an **illusion**. The sun is huge—about 865,000 miles (1.4 million kilometers) around. More than 1 million Earths would fit inside it! It looks small because it's about 93 million miles (150 million km) away.

Fast Fact

The words "sun" and "solar" are only used for our star. Other solar systems are called "planetary systems," and things relating to other stars are called "stellar."

The sun is made up of several layers. The core is the hottest part at more than 27 million degrees Fahrenheit (15 million degrees Celsius). The temperature is much cooler on the surface, but it's still very hot from our point of view—about 10,000°F (5,500°C).

This picture shows what the sun looks like in space.

Fast Fact

The sun is a medium-sized star. There are many stars up to 100 times larger in other parts of space.

Of course, no one can visit the sun to measure how hot it is. Scientists use special tools to figure out what we know about the sun. One of these tools is called a spectrometer. It measures light. Different **elements** give off different colors of light. This is how we know that the sun and other stars are made of hydrogen and helium. Different temperatures also give off different colors, which tells us how hot the sun's surface is. Scientists used math to figure out the core's temperature, but we don't know the exact number for sure.

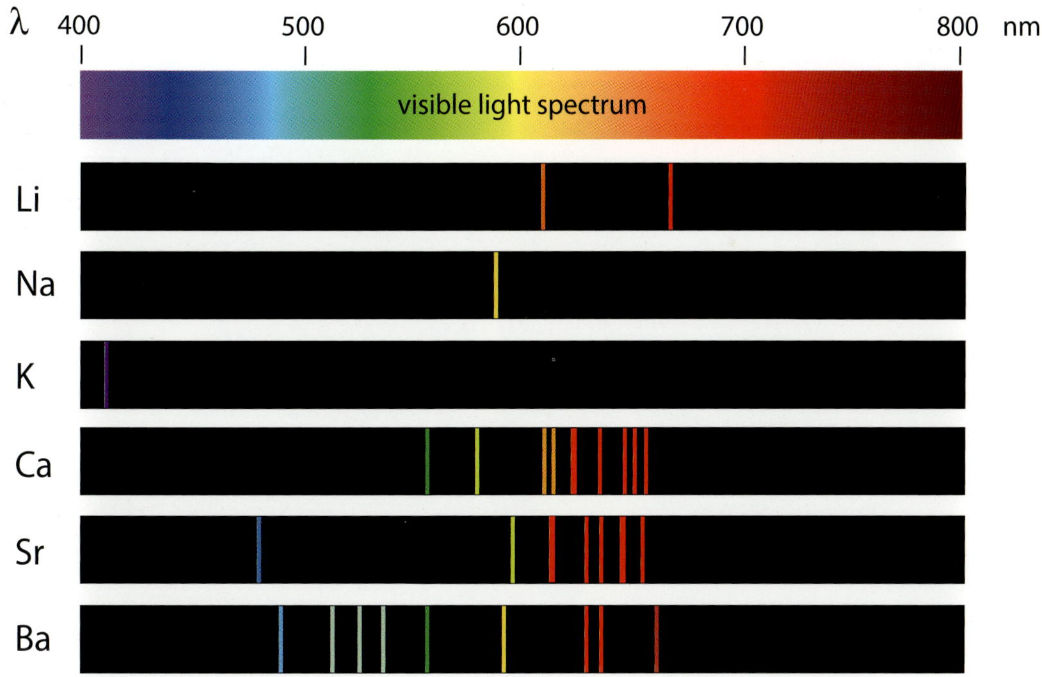

This picture shows the colors, or spectra, of different elements.

SOLAR WIND

The sun is constantly giving off **particles** that travel through space. This is called solar wind. Solar wind can be both good and bad for our planet. It's good because it protects us from other kinds of energy that come from different parts of space. However, because the solar wind is magnetic, large amounts of it can cause problems with cell phones and other electronics on Earth.

Earth also has a magnetic field, called the magnetosphere, surrounding it. This is because of the metals in its core. The magnetosphere pushes solar wind away from us. Without it, the solar wind would have blown away our atmosphere millions of years ago.

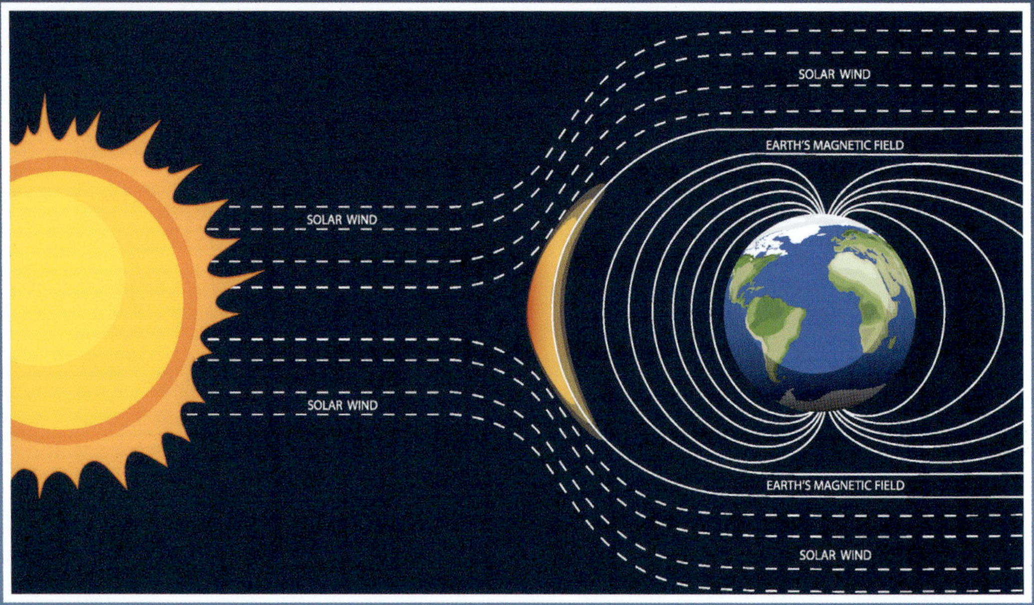

This picture shows how Earth's magnetosphere deflects, or pushes away, the solar wind. This stops the solar wind from getting close enough to the planet to blow away our atmosphere.

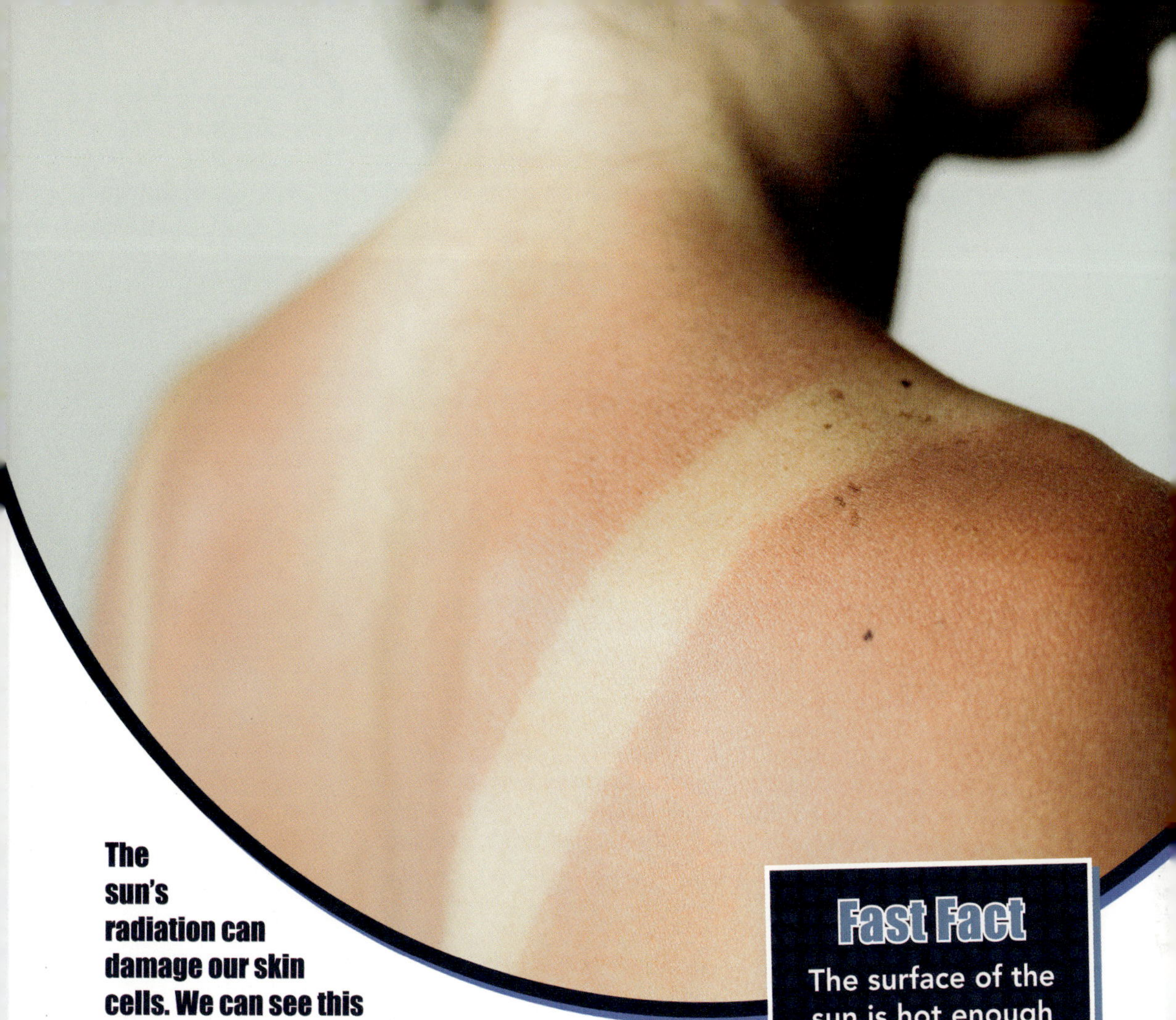

The sun's radiation can damage our skin cells. We can see this damage, and we call it sunburn.

Fast Fact

The surface of the sun is hot enough to boil diamonds.

Solar Radiation

The sun gives off a kind of energy called solar radiation. We call it sunlight, but it's more than just visible light. It includes the heat we feel as well as light we can't see, such as UV light. Earth's atmosphere protects us from a lot of this radiation, but some still gets through. This is why we need to wear sunscreen if we're out in the sun.

A planet, such as Earth, needs to be the exact right distance from its star to support life.

A MEDIUM STAR

The size a star is when it's formed determines how big it will be later in its life cycle. The sun is medium-sized. It's known as a yellow dwarf. When it exits the main sequence, or main part of its life cycle, it will become a red giant, then a white dwarf.

One of our neighboring stars, Betelgeuse, started off much larger than the sun. It exited the main sequence many years ago and became a red supergiant. In about 100,000 years, it will die in an explosion called a supernova. Only the largest stars do this.

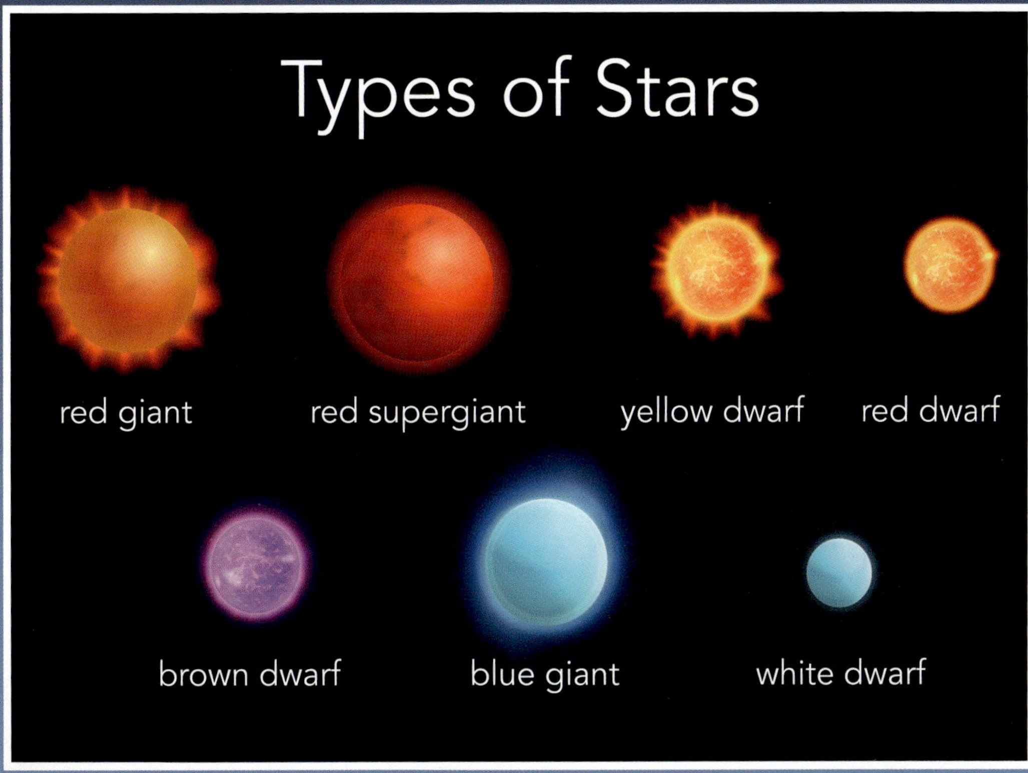

This picture shows different kinds of stars.

Life and Death

When a star is in the part of its life cycle where it's turning hydrogen into helium, astronomers say it's in its "main sequence." The sun is currently in the main sequence stage of its life cycle. It will stay this way for about another 5 billion years, until it runs out of hydrogen. At that point, its core will start to shrink.

Fast Fact

Most stars are very, very hot. However, brown dwarfs tend to be cooler. Scientists have found one brown dwarf that is between -54°F (-48°C) and 9°F (-13°C). This is about the same **temperature** as the North Pole.

When the core shrinks, the outer layers of the sun will expand. It will get big enough to swallow Mercury, Venus, and maybe Earth. Even if Earth doesn't get swallowed, the sun will be too close for anything to live on the planet anymore.

After several billion more years, the sun's core will cave in under the weight of its own gravity. It will turn into a hot, dense ball about the size of Earth. This is called a white dwarf star. The sun's outer layers will be pushed away from the core, and they'll form a new nebula.

The death of our sun may sound scary, but we don't need to worry. No one who's alive today will be around when it happens. By then, humans and animals will likely have found another planet to live on. For now, we can relax and enjoy all the wonderful things the sun gives us every day.

Fast Fact

Main-sequence red dwarf stars are the most common kind of star in the universe.

This photo shows the Helix Nebula, which is the remains of a star that died more than 10,000 years ago.

THINK ABOUT IT!

1. Why do you think ancient people were so fascinated by the sun?

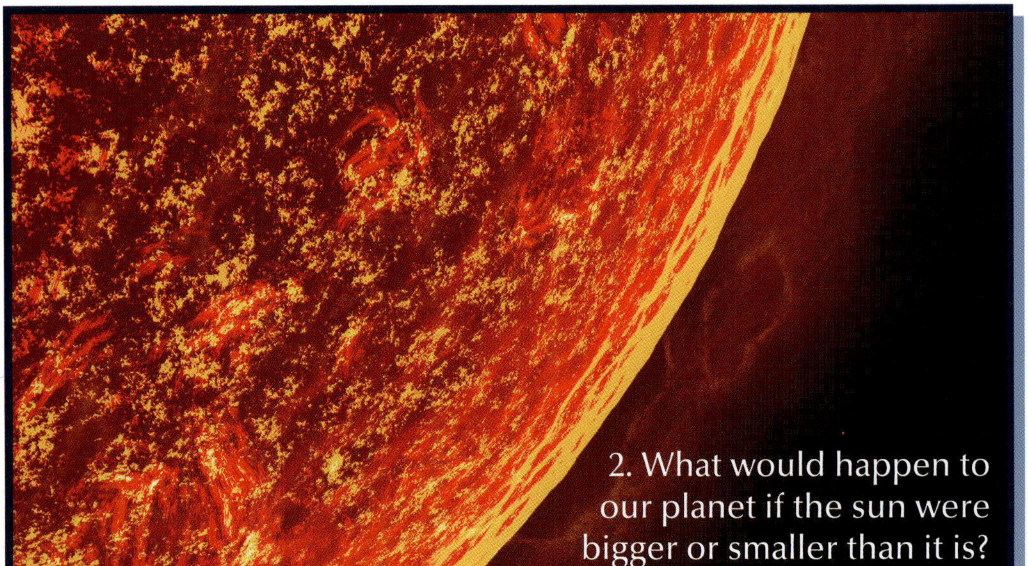

2. What would happen to our planet if the sun were bigger or smaller than it is?

3. How does sunlight help keep our body and mind healthy?

4. What do you think will happen to the outer solar system when the sun becomes a red giant?

INDEX

A DELICATE BALANCE

Chapter Three

The sun is part of an important balancing act that created and maintains life on Earth. If the sun were farther away, Earth would be too cold. If it were closer, Earth would be too hot. The spot we're in is called the sun's **habitable** zone. It's also called the Goldilocks zone because it's "just right."

In the Beginning

The sun and all the planets in the solar system formed 4.5 billion years ago. At first, Earth was covered in hot, liquid rock called magma. It also had a thinner atmosphere, which let more solar radiation hit the ground. Although this would be very bad today, it was just what the early planet needed to spark life. Scientists believe the radiation made **molecules** combine in ways that started to form living things. At the same time, Earth was slowly cooling off.

> **Fast Fact**
>
> Astronomers call the distance between Earth and the sun one astronomical unit (AU). They use this to measure other distances in space. For example, Jupiter is 5.2 AUs from the sun.

About 2.4 billion years ago, tiny living things called cyanobacteria **evolved** in Earth's oceans. They gave off oxygen, which got trapped in the atmosphere. The oxygen allowed more **complex** living things to evolve.

Fast Fact

Today, we know life first evolved in Earth's oceans. However, scientists still don't know exactly where this water came from. Some think water came from space. Others think it rose to the surface as Earth cooled.

Shown here are stromatolites, or "living rocks" made of clumps of dirt as well as living and dead cyanobacteria. They're some of the oldest living things on the planet.

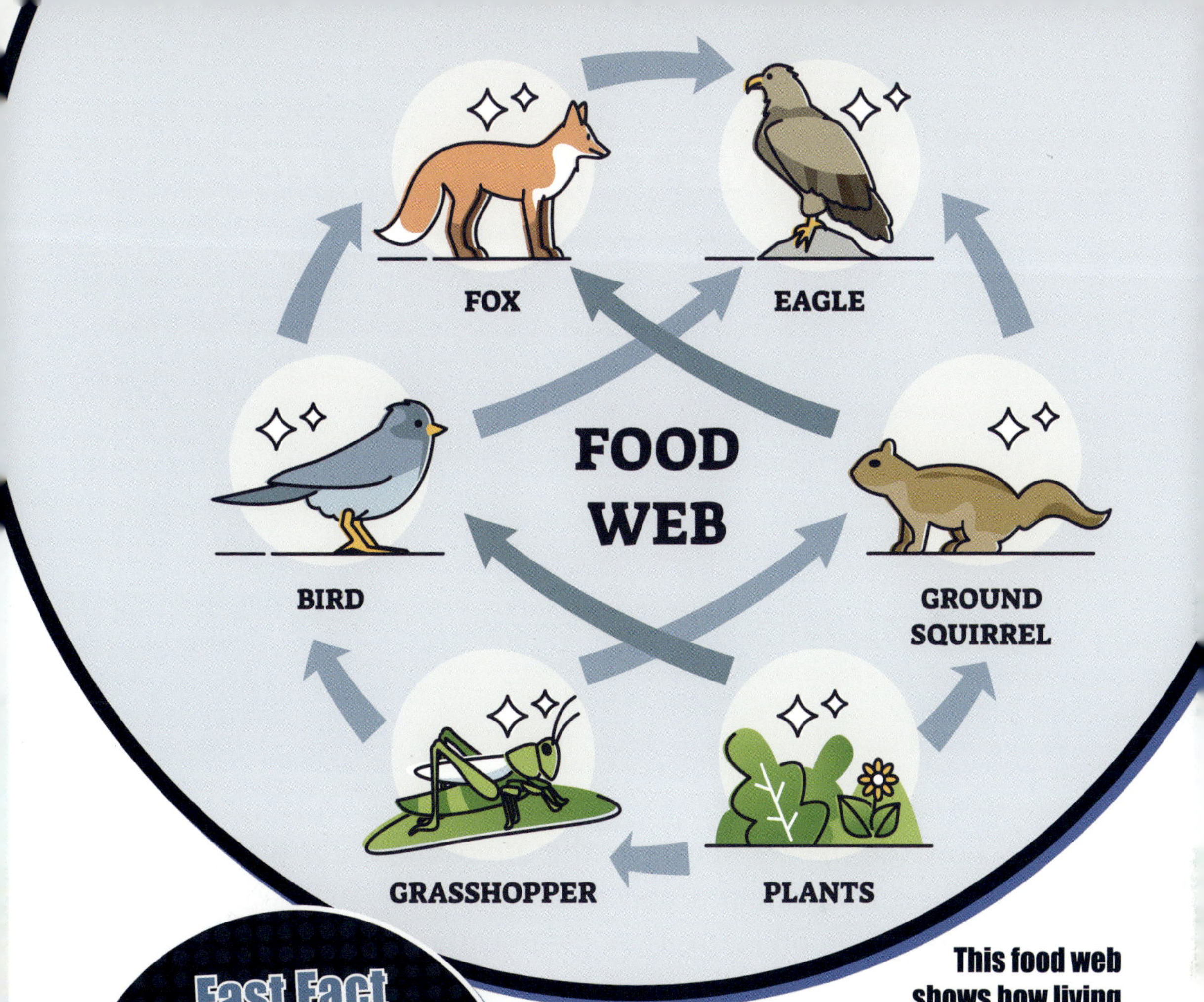

This food web shows how living things are connected based on what they eat. Without the sun, plants would die. Without plants, every other living thing would die.

Fast Fact

Human bodies need sunlight to work right. We don't need it for food the way plants do, but going too long without sun can cause problems in a person's body and mind.

Around 500 million to 700 million years ago, the first plants appeared on land. They used the sun to create food in a process called photosynthesis. They also took in carbon dioxide from the atmosphere and put out more oxygen. Scientists believe

CHANGING TEMPERATURES

The continents are always moving, and in the past, they've moved in ways that blocked the flow of warm water to the poles. When this happens, we get what's known as an ice age. The planet gets very cold, and the snow and ice bounce the sun's heat back into the atmosphere.

The planet gets warmer when more heat is absorbed by the ground and water. Even though big temperature changes can happen naturally, right now the planet is getting warmer because of human actions. This is causing worse storms, more wildfires, and many other problems.

this was how the oxygen levels in the atmosphere got high enough to let animals grow skeletons and **diversify**. Today, plants are still an important part of life on Earth. They clean carbon dioxide from the air and replace it with fresh oxygen. They're also food for many animals, including humans. Plants need the sun, and we need plants!

Sky and Ocean

The sun's heat also has a huge effect on our planet. The sun warms large air masses that move around in the atmosphere. Warm air masses move toward Earth's two poles, which are cold because they get the least direct sunlight. The air masses that form over the poles are also cold for this reason, and they move down toward the equator. When warm and cold air masses meet, we get storms.

Ocean currents behave the same way. The movement of these currents helps keep the planet at a comfortable temperature. Without them, it would be far too hot near the equator and far too cold near the poles.

Iceland, shown below and on the map in red, is very far north. Without ocean currents that bring warm water from the Gulf of Mexico, it would be too cold for anyone to live there.

A nebula where stars are formed is sometimes called a "stellar nursery."

THE SUN'S LIFE CYCLE

Chapter Four

Stars aren't living things, but they do follow a predictable cycle. Astronomers call this the life cycle of a star. Right now, our sun is in the middle of its life cycle, and it will stay that way for a few billion years.

The Birth of a Star

Stars start out as large areas of swirling gas and dust called nebulae. When there's a lot of material in a nebula, it starts to clump up. The gravity of this clump draws more material into it. When it reaches a certain size, the clump starts to rotate while still attracting more material. This turns the clump into a hot, solid ball called a protostar, or "pre-star."

The gravity that pulls a protostar together compresses the hydrogen gas in it. When a gas is compressed, the atoms that make it up bump into each other more often. These bumps produce heat and light. The more the gas compresses, the hotter and **denser** the protostar gets. When the hydrogen atoms get so close that they fuse, or join together, they turn into helium. This is when the protostar becomes a star.

Fast Fact

Most protostars turn into stars, but some never reach that stage. These "failed stars" are called brown dwarf stars.

This picture shows what happens while a protostar is forming.